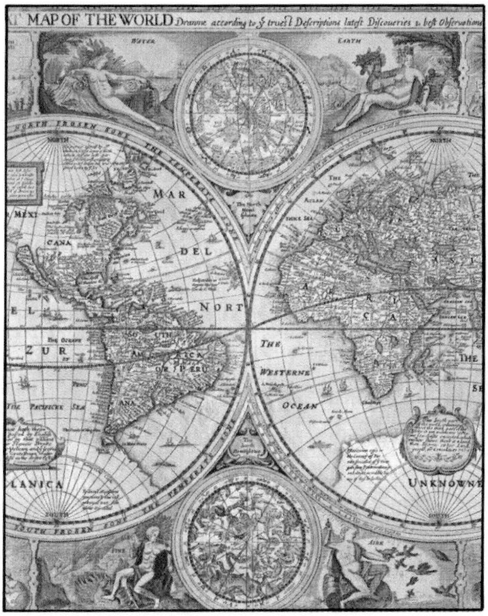
MAP OF THE WORLD